Your Answer...

—THE—
CONNECTION!

Access an Intelligence that provides amazing
resources for success. Without charge!

KIRT G. SALISBURY

"The Connection"

Kirt G. Salisbury

Cover Design and Text Formatting: JDavid Ford & Associates, Dallas, Texas

Printed in the United States of America

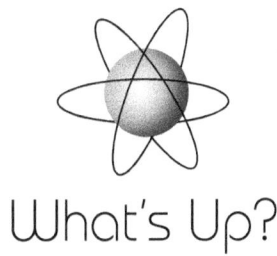

What's Up?

M y desire in writing this book is to encourage and motivate you to think about something that may never have entered your mind. Something that can have a significant impact on your personal life.

I wrote this book to help you to get ahead, become what you want to be, and discover that life can be beautiful, productive and yet peaceful! Let's begin with a step that certainly has helped most, if not all of us, during our lifetimes.

Most people have discovered that in order to make significant advances in their careers or life ... many times it is through the guidance of someone with whom they have connected. Someone who has more knowledge and experience. Someone who has connections with others. Therefore, the right connection can help you advance more quickly and to a greater degree.

This can be an incredibly beneficial step whether you are mature with an established career or a college student looking forward to different opportunities after graduation. Or you may be in a time of repositioning yourself because of today's ever-changing market demands.

Guidance, or the determination of one's direction, is one of our greatest needs, especially in today's world. We are presented with choices way beyond any of the choices generations before us ever had! There are so many options available that many feel it is like playing Russian roulette ... try anything until you find one that works!

This unfortunately is an expensive and time-consuming proposition. Many have chosen career paths only to discover after graduation that this was not the best one either for themselves or for the ever-changing marketplace. Or it may be a good choice for now ... but as you get older and technology replaces more and more hands-on skills ... will it be too late to make a career change then?

Don't misunderstand. I strongly recommend repositioning one's self when necessary for changes in life style or in order to qualify for a different career or job opening.

Unfortunately, there are many talented people who could share their knowledge and experience with others but because of self-centeredness, greed, fear or paranoia, they keep everything to themselves because they are more interested in their own personal advancement. Many times at the expense of others!

While there is a competitive spirit in the business world that at times makes this a necessity, a problem still exists. With whom do you have the confidence to connect? Have you ever found someone who mentored you out of the goodness of his or her heart? If so you are quite fortunate! In days past that is how most of the trades were taught; apprentices were placed under a skilled person to learn exactly how to do something correctly. They received both knowledge and experience at the same time, providing them with a great advantage.

Today we seem to prefer the opposite ... attend college for the knowledge, then after graduation try to find a company willing to hire and train you.

As a former senior executive, I have had people with Ph.D.s working under me. Not because I knew more than they, but because they could not find openings in their career specialty.

So the question is ... where do you find the guidance that will help you make the best career or business choices when you are not certain what the future holds?

We live in a dynamic world of technology that produces change as never before. How do you keep up with it so you are not left "holding the bag" as the old saying goes? Are you that intelligent on your own? If so ... you should have no problem being an exceptional choice for any employer.

These days it is not unusual for college graduates and returning military personnel to be unable to find employment. Career redevelopment may be in order, but at what expense? What do you choose and where do you go? What can you afford? It's at times like these that a local "prophet" would be nice to know!

Most of us, at one time or another, have become anxious facing such major decisions. The wrong decision can lead to bankruptcy. Thoughts of suicide come to many when they feel there is no answer and they are stuck ... hopeless!

How about life itself? When your life goes off the rails, when your wheels seem to fall off and everything that you do goes wrong ... what do you do? Where do you turn? When people give you different answers whom do you believe?

Who has the wisdom to give you not only direction and guidance but also to share their connections with others, connections that could take you a giant step ahead? I found out a long time ago that I can't figure out everything on my own. I need the help of others as well!

What is the "Solution to Success?"

Over the years, I have placed my trust in many people whom I admired or who were very successful in what they did. But I soon discovered that my trust was in vain. As long as I was of benefit to **them**, things were OK, but the minute that changed, their interest in assisting me was over and I was placed on their list of "friendly expendables!"

What I have experienced in my lifetime is the basis of this book. I will attempt to refer you to a connection that you can trust and rely on to a greater degree than anyone I have ever known or met. You may not find this person listed in the Forbes Top 100 however.

Let's start with science and technology. You cannot get around it. In one way or another most of the things we are interested in or we are doing, will in some way involve science.

Science is the basis for all things from basic matter to advanced intelligence. Just what is science? One way to describe it is: "Science is man's discovery of the foundational laws of physics and engineering which determine how things in the universe operate." What we call inventions are actually applications of these existing laws and principles. Man is simply learning to apply them to products and services that facilitate our needs.

Unfortunately, people have long insisted that science must be kept separate from religion. Some reason that science is the invention of man and religion is the truth from God.

In reality both science and religion continue to search for answers such as why we are on planet Earth ... why we are alive ... is there any connection between science, astronomy, technology and the God of creation ... if any?

We have no problem recognizing scientific discoveries but unfortunately this separation has made religion more mysterious and invisible than realistic. Many people who consider themselves intelligent and logical have found both religion and God hard to comprehend and accept.

We also are accustomed to think that if there is a beginning to something, there must be an end. If there is one side there must be an opposite side. If there is an East ... there must be a West. If there is hot ... there must be cold. If there is a positive ... there must be a negative. It even exists in human beings. Humans are born, then they die.

Is this true of everything? To find out, we must go back to the very beginning of life!

How and where did everything begin?

It all began with the creation of basic matter. What is matter? The Merriam Webster dictionary explains it this way:

"The substance of which a physical object is formed ... material substance that occupies space, has mass, and is composed predominantly of atoms consisting of protons, neutrons, and electrons, that constitutes the observable universe and that is interconvertible with energy."

All matter is composed of atoms. Every physical

item that exists on earth and in space from the atmosphere to the ground, from vegetation to diamonds, from our bodies to man-made items all consist of trillions of minute particles called atoms.

Billions of atoms compressed together constitute a physical substance that in most cases we can touch and hold, despite their variations in size and weight. Much of the matter in today's world is what we call solid material. But is it?

What seems to be solid material to us is really comprised of billions of these atomic particles tightly and strongly compressed together. Atoms are so minute that they are not visible to the naked eye. They can only be detected or "seen" with the use of scientific equipment.

Science has concluded that everything we see is composed of things we cannot see.

These invisible particles are made up of electrons and protons which are simply positive and negative charges of electricity. Each atom however, has a dense center called the nucleus which contains neutrons that are neutral and have no electrical charge, and protons which have positive charges.

Interestingly ... since the protons inside the nucleus are all positive charges, they should repel each other and simply scatter. This is a law of physics and electricity! However ... they do not. There exists a binding force, not yet fully understood, holding them together. What makes them stick together like this is one of the many mysterious laws of physics.

What if the law of physics could be overridden by The Force that holds all matter together? Who might that be?

Creator God?

    While their actual form is disputed, for the most part the accepted representation is this basic diagram:

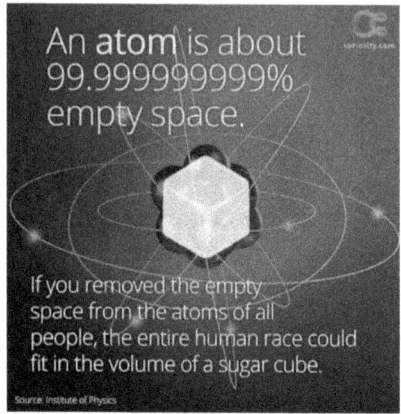

    This is how infinitesimal they are: "A single atom is about 99.999999999% empty space. If you removed the empty space from the atoms of all people, the entire human race could fit in the volume of a sugar cube." (Institute of Physics).

    The study of such particles is called nanotechnology. Much research is underway experimenting with the manipulation of matter at the sub-atomic, atomic, and molecular levels to create materials with new and remarkable properties ... many which never existed before.

    No one has to tell you what is solid. But is that which is solid really solid? If you were to pound your fist on the top of your desk, you would consider that as heresy. But let's take a look at a basic, common item: a single piece of graphite lead from a wood pencil. Looking at it no one would consider it to be anything but solid ... certainly not like steel ... but at least solid enough to hold its own form.

However, looking at the end of a tiny graphite pencil lead as seen through a powerful microscope, it almost resembles the end of a cigarette. It has lots of openings and holes. If we were able to examine other solid materials we would find similar construction. Yes, even steel!

The difference with steel is the density of the material. Its atoms are extremely tight and firmly bound together. The same is also true of our bodies. They contain solid bone matter as well as soft tissue. If this mysterious law of bonding of the atoms were to suddenly be released ... we would collapse into a mound of extremely fine powder!

There is an old fable that I remember as a kid that went this way. An old man in Europe sold vegetables that he grew in his garden. He hauled them to market on his old wooden cart. Concerned that if his cart ever broke down, he would no longer have an income, he asked God, "Please do not let my cart break down until every part of it fails at the same time."

In due season the old man died. In front of his tiny house where his cart always stood was a small mound of dust! Everything fell apart at the same time!

If everything is composed of atoms, what holds them together?

Scientists believe they have already discovered the "God particle." Not attempting to understand this totally but if God created these minute particles that form anything that exists, might it not be logical that He has the power to hold them together?

"For by Him all things were created: things in heaven and on earth, visible and invisible, whether thrones or powers or rulers or authorities; all things were created by him and for him. He is before all things, and in Him all things hold together." (Colossians 1: 16-17 KJV) To me, God is saying that He is the force that holds all atomic structures together!

Early in my engineering career I was taught that everything was in analog form. This meant there was no interruption of flow or inconsistency of energy or materials.

Everything was in a continual form much like the flow of a river. Like the flow of direct current from a car battery. That theory was applied to solid materials as well.

Later came the invention of the digital form of data and molecular construction. It is not continuous but is broken up into minute bits of digital data. For a quick example, look closely at a newspaper photograph with a small magnifying glass and you will see the dot structure of that photo.

Yet from a distance our minds put it all together as one solid picture. Our television sets ... music on CDs ... our cell phones ... computers, etc. We consider this new digital form of energy to be a recent technological breakthrough. But we humans are a Johnny-come-lately because Creator God, upon creating the atom, created everything in digital form from the very beginning!

The question then becomes ... if God created the atom that comprises all matter, every living creature, our human bodies, the universe, planet Earth, don't you think it would be a good thing to know who Creator God is and what plans He has in store for you?

Again, many feel that science and God are two entirely

separate entities. But are they really?
God IS science and technology!

Scientists may not be inventing new things as much as they are discovering how Creator God has designed everything with laws of physics and nature. From this, man has created new materials, equipment, food, chemicals and medical breakthroughs.

Many believe there really is no God because science knows how our universe came into being. However, in recent years, the professional scientific community has begun to realize that science without the God factor does not provide the complete answer.

Many are now admitting that before they included the God factor into their equations, their research was based more on speculation and often resulted in questionable conclusions.

"Science without religion (God) is lame, religion without science is blind." —Albert Einstein

Quoting a high-level scientist at a recent world scientific convention: "As the truth is revealed about what is going on in our skies and universe, we may have to rewrite history. We scientists know that we can no longer limit our thinking, actions and the consequences to planet Earth! Everything we do affects not only us but the entire cosmos!"

Can you believe that God's creative knowledge is available to us either by scientific research or divine revelation? If you question this, check this out: It was God who gave Noah the forewarning of an impending world flood (an event that had not occurred since the time of Adam and Eve) as well as exact specifications on the building of the ark, to save his family and the animal species from death. This was not to be the last input from Creator God to man however ... the list goes on until this very day.

Creator God is not anti-science! Be careful that you do not believe in science more than God. He provided us with brains and intelligence to learn and comprehend the things we need on earth. The choice of how to use God's technology is really up to man. God inspires man with inventions and answers to scientific questions for the good of others. But self-centered man and especially those under the influence of the evil one, Lucifer/Satan, will use this information for selfish purposes.

If it is possible to receive direction and information from Creator God ... how do we receive it? This is THE CONNECTION!

Life is more exciting when you think on a larger scale! Do you think of life as being either on Earth or in Heaven? Most do. The question we must ask is: "Is life really confined to only Earth and Heaven?" Are we thinking too small? If the universe contains billions of galaxies and each galaxy contains billions of stars and planets, what is the purpose of all of this?

Was there a cosmic message that may have, in time, been lost? It would seem that in ancient times, there was a great interest in discovering what the stars were all about. Did someone create a message written in the stars and constellations? Could it be a profound message that told a prophetic message to all mankind from Creator God?

Creator God has always attempted to communicate with mankind in one way or another. "And manifold God made great lights to shine in the heavens. And these lights were for signs, seasons, days and years." (Genesis 1:14-16 KJV)

From the beginning, God posted the message of life in the stars, constellations, and Zodiac patterns. It told the story of creation and the future of mankind as only God could tell it. It was for the entire world to see and comprehend. No one could miss it!

Looking for a fascinating book? "The Witness of the Stars" by E.W. Bullinger (Kregal Publications) provides the complete story. However, over time and with man's interpretations this message has unfortunately been compromised and reinterpreted as mythology and astrology.

Planet Earth: an exceptional planet or just one of many that could support complex life?

On record are some of today's foremost astronomers and astrophysicists who have concluded that among the billions of galaxies in the universe, there is none like the Milky Way, the galaxy that contains Earths's solar system.

Scientists estimate that it has required more than 4 billion years for our galaxy and earth to produce the various molecular structures required for human life support.

Most galaxies appear as extremely large disc-shaped masses in the heavens filled with stars, planets, gasses and other phenomena difficult to explain. However, the Milky Way has a vast hole ... a transparent circular opening that circumvents the space where our solar system resides. Research from NASA's Kepler Space Telescope and the Hubble Telescope are confirming facts about our galaxy that have not been known previously. Scientists have concluded these facts by observing our sun, moon and Earth. They are precisely positioned in their orbits in a way that produces an atmosphere far different from any of the other planets.

Only Earth enjoys a protective shield from the intense radiation of the sun so that there is a thin layer of atmospheric gasses that sustains life on Earth for plants, animals and human beings.

The precise positioning of our solar system within the Milky Way also gives us a clear, visible opening surrounding Earth that allows Earth to be totally and clearly visible from Space. This allows us to also clearly observe the entire universe!

Big Bang Theory

Why would you think all of this was created by one accidental Cosmic Big Bang?

Think on this: Our atmosphere is clear, our moon is just the right size and distance from the Earth and its gravity stabilizes the Earth's rotation. Our position in our galaxy is precisely aligned. Our sun with its precise mass and composition and the Earth's axial tilt perpendicular to the plane in which the sun orbits is 23.5 degrees off axis. All of this combines to give us the environment and atmosphere that allows us to enjoy life as we do!

This was not by accident... it was by design! "In the beginning God created the heavens and the earth." (Genesis 1:1 KJV) Creator God is not some myth or spooky wind floating around that comes and goes without any concern for the people He created. He is real. Our lives which we are privileged to live are but a stepping stone to the future.

One of the most profound astronomical discoveries has recently been highlighted in a television special. This program, aired not long ago, is titled "The Privileged Planet." Two scientists Jay W. Richards and Guillermo Gonzalez produced the show. Both men hold a Ph.D. and specialize in astronomy. A fascinating and beautiful DVD by the same name is available. It's literally out of this world.

This program is not only spectacular in its beauty and

coverage of the universe, it also shows how planet Earth is positioned in such a unique place for man's benefit. You cannot help but recognize the complexity of Intelligent Design that went into our universe!

"All things work together for those who love God ... (Romans 8:28 KJV) Actually, ALL things work together in Creator God's universe.

Nikola Tesla was one of the world's greatest scientists

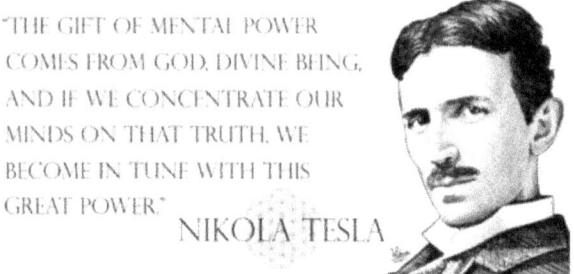

"THE GIFT OF MENTAL POWER COMES FROM GOD, DIVINE BEING, AND IF WE CONCENTRATE OUR MINDS ON THAT TRUTH, WE BECOME IN TUNE WITH THIS GREAT POWER."
NIKOLA TESLA

who ever existed! Incidentally, he invented how to make electricity travel long distances. It was Tesla who invented this rather than Thomas Edison who took credit for it. Other than scientists, few know about this man, but to a degree his knowledge remains and baffles scientists today!

Is there a purpose for Planet Earth?

In the very beginning ... billions of years ago, Creator God, the Supreme Spirit Being who always was and now is and will forever be, began His kingdom in the spirit realm. That is what heaven is all about ... beings who are alive and active but do not have physical bodies. We know them mostly as angels and as millions of human beings who died and were connected to Creator God.

Is it difficult for you to envision an invisible but

functioning world located far up in the north of the Universe? This heavenly Kingdom of God was and still is a perfect place. No evil, crime, self-centeredness, greed, hate or poverty. All spirit beings there are totally submitted to Creator God and desire nothing else.

Then in some point in star-time, God created the stars and planets. These apparently were the first physical objects that ever existed. Once again, everything Creator God plans and does is for a specific purpose. His plan was to create a physical world that would be inhabited by physical beings so they could experience that aspect of God's creation, love and goodness.

Is that why God created planet Earth as a specific place for mankind to live, multiply, grow physically, intellectually and spiritually? Once you understand all this ... in reality, we owe everything to God!

Is being spiritual the same thing as being religious?

Actually they are not the same. We make too much out of being spiritual when it simply means to recognize Creator God and His Son, appreciate all He has done and provided for us and to submit our lives to Him.

Most people seem to be more preoccupied with religion or the reasonings of God than a relationship with God. Religion can keep you in a box of belief systems that may not even be Biblical. In actuality, religion is anything man creates that provides a set of beliefs, rules, religious laws, routines, repetitions and reasonings that are designed to satisfy the mind, logic and emotions of man into believing that in doing these things, he is acceptable to God!

Religion presents a rigorous routine with rules and

regulations invented and instituted primarily by man, often in an effort to convince people that they can earn their way to God!

I hope that you have come to realize that there is a vast difference between religion, the exercise of man to earn favor with God by what he does, and a personal relationship with the living God who is responsible for everything we have on earth today! And Who, by the way, provides us with many things we don't even deserve!

Once you have made The Connection, learning to become quiet and sensitive to your inner spirit, you will in time learn how God speaks to you. The way He speaks to you may be different than how He speaks to others. There is a vast difference between religion and knowing Creator God personally!

You may be thinking, "Yes, but the Bible gives instructions on how we are to live and interact with others." This is true. However, you must keep this in mind. The Bible is the inerrant (absolutely true) Word of God. God Himself inspired men to write it as they <u>understood the languages and word meanings of their day</u>.

The Bible is a true Word from God written by men whose every word was inspired by God. But since that time the Bible has been interpreted by man with his human reasoning!

As you can see, these individual interpretations of the scriptures of the Bible is the reason we have so many different churches, denominations, various dogmatic theologies and even cults.

Did you know that of all the books ever written that

offer advice and guidance for life, only the words of the Bible come directly from God and speak directly to man? It is not considered intellectual suicide to believe in the truth, even if you do not completely understand it.

The Bible is not just an historical storybook. While many find parts of it difficult to understand, it is a guide book to the plans and purposes of Creator God. Not only for the entire universe and Earth but each individual as well!

Earth... a perfect place for mankind?

What happened? It certainly isn't a perfect and peaceful place today! For the most part it seems to become more challenging and evil each year. I agree with many that we are living during a brewing world storm that could end in chaos and destruction of unfathomed proportions!

For now, let's go back to the very beginning of life on planet Earth. There is a multiplicity of theories and opinions of when Earth was created and when man was first placed here. This debate will probably continue until we find out from Creator God Himself!

The book of Genesis in the Bible is probably the most misunderstood, debated and frustrating book of all time!

Why? One reason is when Creator God inspired the writer of Genesis, His purpose was not to give an elaborate account of every aspect of His design and purpose for the entire universe or the actual beginning of creation, but to present to us a few fundamental facts.

The total creation story of Earth is told in only 31 verses in only one book of the Bible ... a story that took place over millions, if not billions of years! Yet 38 entire books of the Bible were set aside to give an account of God and the history of man on Earth. In God's eyes it was more important that we understand our relationship with Him as He relates to us on Earth, than a mire of technical and scientific details on the phases of the entire universe and its creation.

Many Bible believers strongly defend their belief in the inerrancy or absolute truth of the Bible. They believe it says that God created planet Earth and then the first humans, Adam and Eve, to inhabit it. To them that means that planet Earth is only about 6,000 years old ... not billions of years old as science tells us.

Genesis tells us that God created the earth in six days. Yet in Genesis verse 1 it says the earth was void and in chaos ... in ruin! God certainly did not create it this way! This then was God's recreation of earth.

This may be hard to comprehend but there were actually two creations of Earth and even much of the universe. The first one is what God created in the very beginning. The second one is the one the Bible gives history to that was the recreation of Earth after its massive destruction millions of years prior.

I find it interesting that there is documented archeological proof that the Chumash tribe of Native Americans settled and lived in California near the Santa Barbara coast some 13,000 years ago! And this is only one of thousands of documented examples!

The fossils, bones of pre-Adamic dinosaurs and multiple sedimentary layers prove that Earth has been around for millions of years. I wonder how people who do not believe this fit this into their reasoning when they visit museums and see examples of archeological and geological proof? Do they conclude that God is playing a joke on humanity?

So when did the "Very Beginning" actually began? No one really knows! The answer to that would challenge the best intellectual minds of today.

But if geological and archeological research is true, then what was going on in the pre-Adamic world before Adam and Eve? If there existed mammoth prehistoric animals ... why could not man have existed then as well?

The truth is ... many believe they did. While certain scientific and religious disciplines still repudiate this, footprints, bones and even strands of human hair of prehistoric man along with prehistoric animals have been found.

Who was this pre-Adamic man?

Primitive man once existed, called Neanderthals or Cro-Magnon. They were human yet different and have no DNA biological connection with humans since Adam and Eve. Is it possible that a form of man existed before Adam and Eve without violating the inerrancy of the Bible?

It would seem that the most important Biblical message to mankind from God was not so much about the ancient universe as it was about the activities on Earth from the time He created a **new form** of man.

Why was a new form of man needed? If you recall in Genesis 1: 26 it says that God created this new man in His image. What does it mean "In His Image?" While the image of God has many, many different facets, God also is known as the Trinity.

Scholars of Scripture know that Trinity means God the Father, God the Son Jesus and God the Holy Spirit. While three entities, they are all One God in every way.

When He created mankind in His image He made man, Adam and Eve, as a triune being consisting of a body (our physical frame) a soul (our mental capabilities that program our body functions) and a spirit (the part of each of us that is unique, eternal and never dies).

Can we communicate with God?

It is from our inner spirits that we hear from God.

" **B**ut there is [vital force] a spirit [of intelligence] in man, and the breath [inspiration] of the Almighty gives men understanding." (Job 32:8 AT)

Just in case you do not believe that Creator God is relevant or practical, think about how our human bodies and the bodies of animals are designed and constructed. Lungs to oxygenate ... stomachs to digest our food for energy and growth ... intestines to separate nutrients from waste ... and the list goes on. Nothing missing in creative design here!

"And the very God of peace sanctify you wholly; and I pray God your whole <u>spirit</u> and <u>soul</u> and <u>body</u> be preserved blameless unto the coming of our Lord Jesus Christ." (I Thessalonians 5:23 KJV)

Keep in mind that only God can create spirit beings! The reason they are eternal is that once created, they never die nor can they be killed! "But there is a spirit in man: and the inspiration of the Almighty giveth them understanding." (Job 32:8 KJV) "Then shall the dust return to the earth as it was: and the spirit shall return unto God who gave it." (Ecclesiastes 12:7 KJV)

This may surprise you but when God announced to the angelic world that He was about to create a new man in His image they could hardly believe it! They spoke out, "Now man will be like us and never die!"

Why would this surprise the angels? Because humans on planet Earth up to this time consisted of a body, a soul but no eternal spirit part. So when they died, these prehistoric beings that some call Neanderthals, or cave men, simply existed no longer.

Don't misunderstand what I am saying here. I am in no way implying that humans and animals are the same or that man came from the animal species! But in the pre-Adamic era man was created in similar physiological structure as the animals. They lived a full life but when death of the body occurred, they existed no more.

Why would God do this?

To most, this is incomprehensible: a war in Heaven? The most perfect place in the Universe? You are joking, of course. No? How could that really be? God would never allow such an event!

How many times have you been told that Adam and Eve are the ones that brought sin and disobedience into the world? And because of this everyone born since are born in sin! Yes, the Bible says this.

Well, it is true. They did disobey God in the Garden of Eden, so therefore they were the first new humans to do so and thus pass this on to the entire race. But they were not the initiators of sin. That title goes to the most intellectual, beautiful, spirit empowered angel with the highest authority in all of Heaven ...Lucifer!

Who the hell is Lucifer?

I t may be hard to comprehend but it is true and what he did is still the root cause of everything that is against God to this day! Lucifer's sins were pride and rebellion!

Who gave Lucifer such a dominant role?

The archangel Lucifer was created by God to be one of the most intelligent, creative and responsible archangels which He could totally trust. Lucifer was second in command to God Himself! He was the most intelligent and beautiful of all the created beings in the entire universe!

God assigned Lucifer authority over all the angels in Heaven. He was also the leader of worship at the very throne of God. He was there when God created the universe, Earth and man. He learned everything he knew from God!

Wait a moment ... isn't Lucifer the one that is also called Satan? How could someone so evil be that intelligent? Unfortunately, many think that Lucifer/Satan never really existed or is now powerless to hinder or affect anyone. Never forget. He is far more powerful and shrewd than you think! And he is still alive and active today!

He is not the funny little devil in the red suit with horns and a pitchfork! His name was not changed from Lucifer to Satan until he rebelled against Creator God. And that was billions of years ago!

In the meantime, after God created planet Earth and the pre-Adamic people to begin populating it, He chose the best archangel He had to have authority over its growth and population. This assignment gave Lucifer/Satan the freedom to exercise every God-given gift he had and to prove to everyone how creative he was.

Very few understand this position of power and authority Lucifer was given by God in the very beginning! (Ezekiel 28:13-17 KJV)

Lucifer/Satan had such spirit power, even to create on Earth and in the animal kingdom! He did such a good job of it that he surmised that he could do everything God could do. At that very moment ... sin and disobedience to God entered his mind ... and pride!

He announced to the millions of angels that God placed under him to aid in the development of planet Earth, "I will ascend into heaven, I will exalt my throne above the stars of God: I will sit also upon the mount of the congregation in the sides of the north: I will be like the most High" (Isaiah 14:13-14 KJV)

Lucifer boasted that he would replace God in Heaven and on Earth!

What was this all about? Control and authority over the Kingdom of God both in Heaven and on Earth! It was a daring and presumptive step by Lucifer. He had much to lose if he did not win but his pride told him he could actually conquer God and His Heavenly Kingdom!!

To make a long story short, Lucifer and his angels went up to heaven and fought against God's army headed up by the archangel Michael and his angelic forces and lost! Such massive defeat and humiliation had never happened before. This is when God changed Lucifer's name to Satan ... the evil one!

From this point forward, Lucifer/Satan did everything he could to discredit God. With his powers of deception, his propaganda program began spreading his lies over all the Earth and continues to this day! Lies such as the following: God is not good! God really doesn't even exist! God is legalistic and certainly does not love you!

Could this be why many have problems believing God? He is not the God that is cruel, legalistic and judgmental that many have been taught but a loving God of compassion and mercy. However the big lie keeps being propagated by Lucifer/Satan and those who do not know God causing many to miss a beautiful life of eternity with Creator God in the new worlds to come!

The Sin Issue was settled long ago.

While Religion appears to be focused on condemning us with the sin issue, that really isn't the issue at all. Yes, sin separates us from the presence of God but God and His Son settled the sin issue many years ago for every living being on the face of the earth, both now and in the future! (John 3:16 KJV)

So from that point on, we are not condemned by God for our sins, but rather for not accepting the free gift from God; Salvation by faith in the redemptive work of Jesus Christ, by His death on the cross!

The first new form of humans on earth Adam and Eve, our parents if you will, did disobey God in the Garden of Eden. Therefore they can be considered as the first to sin among mankind. But remember this -- it was Lucifer who introduced the first-ever sin into the universe and our world! Then he deceived Adam and Eve to follow in his footsteps!

As a result of this action on Lucifer/Satan's part, God decided to give authority of planet Earth over to the new mankind that was created in His image who would inhabit Earth. Man would be commissioned to cultivate it, prosper it, populate it and as much as possible, create a physical version of Heaven on Earth.

However, God did not totally eliminate Lucifer/Satan but instead restricted his ability to destroy or kill mankind. So the primary work of Lucifer/Satan has been to deceive us! His best deceptions are as follows:

- There is no God
- Even Satan himself does not exist
- God is evil and does not love anyone
- God only wants to punish you and send you to Hell
- Satan is good and will give you power over others
- God doesn't want you to have any fun in life
- Satan will give you all of your heart's desires, no matter what they are

And the list goes on.

The only problem in receiving what Lucifer/Satan gives you is that he demands a huge price for his favors. Your soul! He wants you to share in the price he will be paying in Hell for all of eternity! How is that for compassion?

So if you want to blame anyone for all the evils of this world, don't blame people. Blame the source! It would appear then that the inspiration of evil-minded people comes from the evil spirit world of Lucifer/Satan and his demoniacs!

Where do we come in?

Adam and Eve came into being in the magnificent Garden of Eden. They, as us, have been given free will by Creator God. Our own will to choose and make decisions. In essence, God has put the decisions and direction of not only our own lives in our hands but the entire world as well! A great personal gift to be sure, but if not used responsibly it also can lead to some horrific consequences!

But God did not leave us empty handed. "If any of you

lack wisdom, let him ask of God," (James 1:5 KJV)

Let me clarify the misunderstanding about God and money. Some believe that God is good and money is evil! That is farthest from the truth! In the Bible God speaks of us becoming wise investors in many different ways, another reason why He gave us brains! Poverty does not equal spirituality.

While one of the problems with attaining great wealth is that it can produce greed, another symptom is that it gives a person power and control! With unlimited funds you can virtually buy yourself through any sickness, legal problems, disasters, losses, etc. This provides one with a great sense of self-sufficiency that implies they have no need for God!

I just watched a tiny young wren begin building a nest for her first time. It was like watching a skilled craftsman ... intricate and complicated, yet she did it with skill! She did not learn it from her mother, as this little wren was born in the nest her parents built. Who taught her?

Yes, some behaviors and patterns are instinctual, but where did this come from? Her parents' DNA? Her genetics? Or was she created this way by Creator God? It is my opinion that Creator God implanted a DNA knowledge in all living creatures so they can live, function and reproduce after their kind.

Interesting thing about animals. They know who they are. They know what they are to do in their life. Yes, they do learn from their parents but this knowledge they have is beyond instinctive! Is it possible this could be a parallel to Creator God giving the exact instructions to Noah on the building of an ark that he had never seen? (Genesis 7:14-16 KJV)

Do we really have a purpose?

Creator God enjoys and delights in people ... human beings who reflect His image. He expects us to grow, to learn to use our brains to advance ourselves, others and the world in general. And to demonstrate Godly love and compassion just as He does.

But why did He create us in the first place? Did you know that even though we have been given free will to do anything we wish, God created all of us for His purposes that are for our good? But the choice is still ours!

How do you feel when you hear the word love? Can you relate to it as a wonderful reality, or do you feel like it is a lie? Do you judge it as deception because true love has not been your experience? Is it a feeling that you are leery, even fearful of? As if once you were to ever experience it, it could easily be taken away from you?

If God created perfect love ... love that is unconditional, true, faithful, kind, considerate, without condemnation and always, for you, is it possible that man, assisted by the evil spirit world of Lucifer/Satan, has corrupted this perfect love and replaced it with a humanistic version of love? One that is all emotion. One that is self-seeking, sexually motivated and exploitative beyond measure?

Do you realize that every living thing comes from Creator God? Not Mother Nature. She is only reproducing what God designed.

Look at it this way: "God created music, wine, poetry, sunsets, not man. Who designed the human form in such a way that a kiss could be so delicious ... and more! God is a romantic at heart and fights for every man who loves Him. Eve was the crown of creation. The naked body of a woman is a portion of eternity too great for the eye of man. The business world on the other hand requires a man to be efficient and punctual. Corporate policies and procedures are designed with one aim: to harness a man to the plow and make him produce. Yet the soul of man longs for passion, freedom and life!" (" Wild at Heart" by John Eldredge, NELSON BOOKS, 2001)

A woman's heart should be so hidden in God that a man must seek Him just to find her.

But our top priority purpose is to decide, after experiencing all that God has done for us and given us, if we wish to submit our lives to Him and let Him guide and direct us. Or do we want to do as the popular song sang by Frank Sinatra boasts, "I did it my way!"

What difference does it make which way we choose? If you only knew! God desires a people who love and respect Him and will fit into the great futuristic new World-of-Worlds that He has planned that is coming in the very near future!

But until we have submitted our lives to Him, we will continue to be self-centered, and the strength we enjoy will only be our own. After all, as long as you think you are really capable and are something in and of yourself, what do you need God for?

Are you concerned about safety and protection?

My background is intensely protection oriented. My military and para-law enforcement experiences have prepared me in this way. I have also taught how to survive in the wilderness for many years. My wife used to say "You are not prepared for a national disaster ... you are prepared for Armageddon!"

But I soon discovered that you cannot be prepared for every contigency! There are way too many variables we could face. While we may be prepared for ten incidents there are probably ten more we are not! It's somewhat like playing Russian roulette!

Only God can protect you from everything ... supernaturally! Look at the ecological disasters that are inundating America and the nations of the world, right now! The civil uprisings ... terrorist threats and attacks ... new strains of diseases that are resistant to antibiotics and the return of old diseases that we thought were wiped out years ago!

God has the ability to place a hedge of protection around you that cannot be penetrated by the evil forces of our world. When the great world flood came upon the earth, God gave Noah the plans on how to build an ark. This was their means of escape that no one else had or could even conceive of!

His angels are not only His messengers but they can intervene physically to protect us. I could write a book alone on the many ways angels have protected me over the years! Unfortunately, many who hear of such experiences simply consider it coincidental or just plain luck!

God can also supernaturally intervene on your behalf in ways that would blow your mind! Maybe you are saying, "Well ... God has never done anything like that for me!" How do you know? Have you not experienced situations in your life that defied explanation? Because you did not consider it a miracle from God doesn't mean it wasn't!

But more than that, all of us have no idea how Creator God has supernaturally protected us and our families from danger, over and over again that we simply did not see, yet it happened!

Will evil continue to exist forever?

Very soon God will bring to a climax the evil and deception clouding this world. This is not someone's fantasy. It is a promise from God Himself! It will transpire in a period of only 7 years. It is a time when evil will reign on earth beyond your wildest imagination! It will be far greater than any sci-fi horror movie you have ever seen! It is referred to as the Battle of Armageddon and the time of the Great Tribulation.

This will be a time when every evil, sadistic, anti-God and anti-Christ person and nation on earth will rise up and attempt to take over the entire world, especially the nation of Israel! The first 3 1/2 years will be the greatest deception mankind has ever experienced. It will appear as peace but at the end of this time line, all Hell will break out!

Many think that Lucifer/Satan from the time of his defeat and humiliation in Heaven has been sitting back on his haunches biding his time.

Lucifer/Satan has a legal spiritual right to deceive and control people who have chosen not to submit their lives to Creator God.

In the atmospheres above earth he has been fortifying his demonic armies. Not only the spirit demons but also the humanoid physical beings that he has created. He cannot create human beings made in the image of God however!

His strategy is a terrifying final, global attack against all of humanity.

Serpentine beings ... huge weird creatures that defy imagination. Flying space machines that will send chills down the spine of everyone who witnesses them. It will be a war unto the final cataclysmic end! One-third of the population will be killed.

No one will be able to escape this spiritual and physical battalion of evil, anti-God beings that could soon be released on Earth! Earth literally will be experiencing a Satanic stampede!

Yet millions of people will escape!

Who are they? Those who have chosen to follow God in their lifetime. They will be transported off the face of Earth just prior to the evil horde of demons' release. "It will all happen in a moment, in the twinkling of an eye, when the last trumpet is blown. For there will be a trumpet blast from the sky, and all Christians who have died will suddenly become alive, with new bodies that will never, never die; and then we who are still alive shall suddenly have new bodies too." (I Corinthians 15:52 Living Bible)

Where will they be taken? Up into the heavens to be with God and His Son, Jesus. But it will only be until the great war of Armageddon on Earth is finished. Contrary to what many believe ... the children of God do not remain in Heaven for eternity! Not yet!

God's purpose in this great battle and release of the most evil and sordid human and demonic armies is His Judgment against all who have rejected Him! It is also part of His plan to cleanse the earth from evil influences once and for all! But, while one-third of the people on earth are killed during this time, many will still be alive.

God will then send His Son, Jesus who is the only triune part of God that lives in a human form. And all the saints, the children of God, will come back to earth for a period of one thousand years.

Jesus who has lived in our shoes knows how we humans function. He knows this world like the back of His hand. He will set up His throne in Jerusalem and have complete authority over the nations of the world. Earth will finally be governed by righteousness and truth from this point on.

What will our earthly function be?

We will govern the affairs of this Earth with Jesus and reclaim this world for God by assisting in the restoration of Earth. We will assist God in bringing Earth back to what it was when God originally created it!

Those people who in their lifetime did not choose God yet managed not to be killed and remain on Earth after the Battle of Armageddon will be given the same choice as we have today. Which do you prefer? God's way or your way?

Unfortunately, if they don't chose God's way, they too will, by default, be sent away at death to the inescapable, desolate, hopeless place reserved for the billions of people who have rejected Him since the time of Adam and Eve.

"There is no man that hath power over the spirit to retain the (his) spirit; neither hath he power in the day of death ..." (Ecclesiastes 8:8 KJV)

At the conclusion of the one thousand years the job will be completed. All evil will be bound up in the Bottomless Pit, Hell as it's called, where no part of God will ever exist. Then what?

There is an awesome promise in the Bible in which God attempts to tell us what is in store for us, knowing that we do not have the intellectual capacity to even understand it. So, believe me ...this promise from God is very understated!

"Eye hath not seen, nor ear heard, neither have entered into the heart of man, the things which God hath prepared for them that love him." (I Corinthians 2:9 KJV) "For since the beginning of the world men have not heard, nor perceived by the ear, neither hath the eye seen, O God, beside thee, what he hath prepared for him that waiteth for him." (Isaiah 64:4 KJV)

Would you ever think that Creator God is so limited in His thinking and planning that He is only concerned about the expansion of His truth, grace and love to planet Earth?

If the entire universe is His, the stars, the planets, the cosmos, the billions and billions of space-distance, why would He limit eternal habitation only to Earth? We may be very surprised when we learn what God's future plan holds for us!

I once had the privilege of conducting street television interviews with people in London. When asked "Do you love God?" The most repeated answer I got was, "Why should I? He hasn't done anything for me!" Can you see what a blind statement that is?

But it is not unusual. Many are too busy with personal issues, the complexity of life, concern over the future or the desire to be recognized and entertained to take enough time to even consider what God has done for them.

At any moment you are but a heartbeat away from the termination of your life!

What would you think if one day you opened your appointment book and found it reading: Tuesday 26th 3:00 a.m. -- death? Would this cause you to investigate the possibilities of life after death?

How is this for an interesting question? Do you believe that intelligent spirit-beings actually exist? These are simply beings without bodies. Many people have encountered spirit-beings at some point in their lives. Many have seen or experienced the ministering or rescuing actions of angels.

Others who are enamored with controlling others and have delved into the Satanic arena of spirits have also experienced spirit-beings or fallen angels. These are considered to be in the demonic spirit group. These spirits mostly use deception to accomplish their diabolic plans against mankind.

Creator God is so awesome! Can you believe that when you first call upon Him, and at that point you may not consider that as praying ... He hears you? Then after you make the Connection, God Himself by His Holy Spirit, now dwells and resides within you forever?

Then when you pray you are not attempting to reach Heaven trillions of miles away, you are speaking to the personage of God by the Holy Spirit of God within you.

He knows your every need ... every desire ... every hope ... every cry of your heart even before you tell Him! Awesome, awesome God!

That is how near God becomes to you. No longer will you wonder if God knows you. If He cares about you. If He loves you intimately. Now you have access to Him for guidance and assistance for any area of your life at any time.

God the Holy Spirit is a loving gentleman who lets you exercise your own free will yet He also will warn you of potential harm!

The wisdom of Solomon is quoted in Ecclesiastes 3:1-2, "To everything there is a season, and a time to every purpose under the heavens, a time to get and a time to lose, a time to live and a time to die ..." Here we go again, a beginning and an end. Or is it?

The Bible tells us that, "In the beginning God created the earth," yet man has proclaimed for centuries that, "The end of the world is near!" Why? Because of religious deception, paranoia and fear. When the Bible says, "And then the end shall come ..." it is not referring to the final destruction of the world! Rather an end to an era of evil on earth in the hearts of men and their actions and rebellion against God!

You Have The Power To Choose

**There is a penalty to pay if we reject
God and His free gift of grace.**

Here we go again! The freedom of choice is with our free will! Ask yourself, "Which would I prefer ... living forever in a world of wonder, peace and advanced technology or existing in a world of darkness and remorse in total void, where the presence and goodness of Creator God will never be experienced ever again?"

This new wonder world of God will be unlike anything you have ever experienced in your life. No limitations of your creative mind or physical body. Your supernatural body will be given to you by God Himself in order for you to assist Jesus to install a new order of righteousness in this world, once and for all.

You will not lack anything! There will not be anything you cannot do. You will not need rest or sleep. Nor food and drink. Yet you will be able to enjoy the fellowship of others along with fine dining, anytime you desire. And it will never end!

**Creator God, Father God is your spiritual father ...
believe it or not!**

The chain of command goes like this: you as a son or daughter are under the authority of your earthly father until you are of legal age. Your earthly father is under the authority of Father God for his entire household. However, both of you are directly under the authority of God who created you for all Eternity!

What keeps us from a direct sonship relationship with Father God is contained in a word no one likes to hear: sin! While there is much that is included in that word, it is not just murder, crime, robbery, immorality, as some think. It is simply not living up to a righteous, moral, loving, compassionate standard that God designed us to follow. It's called missing the mark!

The dictionary calls it a standard of performance, right living. Some, however, think they can attain this on their own by being considerate, donating to charity, building orphanages and having people say after you die, "What a good man he was!" But in God's eyes, that doesn't hack it.

The reason is without a change of heart the good deeds we do are simply superficial! In fact, a lot of what is done in the name of Christian charity is simply a facade and a public relations show. Yes, the Bible says that we are to "Love our neighbor as ourselves." (Mark 12:33 KJV) which leads many to question, "Why is this not enough to please God in itself?"

The fact is, in ourselves we cannot please God enough for Him to openly receive us as perfect, holy and worthy of His acceptance. That comes only by believing in the redeeming work of Jesus by His death on the cross.

We in ourselves cannot comprehend how Holy God is and would like us to be. Not in religious ways that we think of as holiness, but in purity of our hearts, our thoughts, our intentions, our motives and our actions. And to think that God knew this when He created mankind!

Why would anyone fear God, if He is who He says He is? Many fear God just as they distrust other people. They really dread transparency! They may even be afraid to face themselves for fear of not being able to handle the truth of who they are or they fear the mental/emotional consequences.

This is an exaggerated lie in their own minds. They prefer to spend their lives in emotional squalor rather than becoming all that God intended them to be!

Many people fear God because they think following His guidance for life means a stiff, straight life style that leaves no room for enjoyment! Not so! In God, life is free, creative, enjoyable and prosperous. However, if you choose to step out of God's guidelines there is a price you will pay, in one way or another, but it's all of your own doing!

Yet while God knew this, He did not leave us to alone. He designed a plan from the beginning of time to have someone accomplish it for us! His most prized possession was His Son Jesus who was to pay whatever price was required by spiritual law to obtain what we could not do ourselves.

The price was the **death** of a **perfect** human being. Believe me … after all the years I have lived on this planet I have yet to meet such a perfect human being … it took a God-man! The name of this God-man is Jesus!

It is a combination of a humbled heart and the realization that Jesus and only Jesus paid the price and opened the way for each of us to personally connect with God the Father. You may not totally understand this, but until you make "The Connection" you are a spiritual orphan away from Creator God, your Heavenly Father. You are in need of adoption!

In the course of our lifetime, many things we are

taught are out of balance to one degree or another. We hear so much about Jesus. "It's all about Jesus!" Well, in a way it is because of the price He paid for us. But there is another reason that many believers overlook.

Without going into the many Old and New Testament parallels, man was never allowed to approach Creator God on his own. There were many conditions and requirements that had to be accomplished in order for man to even be represented to Holy God! But once Jesus paid the price by His death and the shedding of His blood as an atonement for every person on the face of the earth ... this opened the pathway for man to go straight to Creator God Himself.

In other words, you and I do not need a mediator between us and God. We have the access to go to Him directly in prayer! Now, don't make prayer as religious as many do. Prayer is simply talking to God as you would another person. Isn't that how you would create a relationship with anyone? If that doesn't blow you away, I don't know what would! That is absolutely awesome!

Jesus said for us to pray this way: "Our Father which art in Heaven ..." Notice He did not say pray to me! When you begin to learn about Creator God, you are in the process of seeing a much bigger picture of the great eternal plan of Almighty God for all of us!

You do not have to feel unworthy when you talk directly to God! As amazing as this sounds, God sees you in the same state of perfection as He sees Jesus! Jesus is also your representative. He also speaks on your behalf. How good is that?

It is also amazing to me how God seems to have a greater love and concern for the less fortunate than most of us do. He always considers the lowly, the seemingly insignificant, and gives them hope and value in their lives. God also desires that we would do the same.

Are you afraid to admit that you really do not have it all together?

Do you ever look back at your life and see a pattern of chaos in one form or another? The dictionary explains chaos this way: "A state of things in which chance is supreme ... a state of utter confusion."

When you make "the Connection" many new things happen! You also receive a "conversion" which takes your life out of chaos, of going around in the same circles and places you into cosmos. The dictionary explains cosmos as, "An orderly, harmonious, systematic universe ... harmony." You will have a new sense of purpose and direction.

By being taken out of a routine of chaos and placed into cosmos you are now in touch with the greatest loving power the universe has ever known! You will become spiritually alive. You will have inherited life forever with Creator God. How can you top that?

At the same time you will be the recipient of all of the promises and blessings of God as called out in the Bible. All the hurt, shame, paranoia, lies against you, sorrow, low self-esteem, ineptness, hate and injustice you have experienced in your lifetime can be healed by the supernatural power of God.

And you will experience the peace that passes all understanding which the world does not understand, nor can it give to you! God will watch over you and yours ... protect you, guide you, bless and prosper you all the days of your life!

The Connection

**Today, make the Connection that connects
you to the very source of life itself.**

t also makes you a child of God who will inherit all the
promises of God for today and for your eternal future! There
is no cost ... no tribulation ... no penance ... no performance
requirements ... just a surrendered heart.

However, by not committing your life to Creator God,
by default you are unknowingly opening yourself to being
influenced by deception, greed and self-centeredness from the
spirit world of evil that hates God!

No one will be able to say, I didn't know! "For His
invisible attributes, namely, His divine nature, have been
clearly perceived, ever since the creation of the world, in the
things that have been made. So they are without excuse."
(Romans 1:20 KJV)

What does that mean ... commitment?

In today's world the word commitment is not popular!
Many fear the responsibility and effort this word implies. Then
to hear the word covenant that the Bible speaks of can cause
some to break out in a rash!

Let's remove some of the misunderstandings that
seem to cause so many to fear. Submission ... commitment
... covenant relationship ... are not some legalistic forms of

overly-strict judgmental rulership that one step out of line and God sentences you to an eternity in hell!

In reality, God provides us with His wisdom on life to live by, not to make us godly robots or to judge over us every step we take. Creator God established His eternal rules of order in the very beginning. When we stay within His boundaries, we are the freest people on the face of the earth!

Permit me to offer this parable: You decide to join one of the military branches. You swear that you will faithfully defend the United States of America ... be obedient to the commands of your leaders ... learn the procedures and strategies required of a soldier ... be prepared at all times for the unexpected ... and be loyal at all costs ... even to the point of death!

But this is only one side of the commitment. On the other side, the military's commitment to you is to provide you with shelter and housing, food, health services, clothing, transportation, training, weapons and even pay you! Not a bad trade-off, no?

Surrendering your life to God.
Why is this simple step so hard? It is simply our pride!

You are giving Him authority to guide and direct your life not only for today but for all of eternity. To teach you what the Kingdom of God is all about ... protect you with His mighty angels from all perils ... provide for all your needs ... prosper you with the use of His wisdom ... train you as a soldier in spiritual warfare to defeat the enemy of our souls ... and inspire and speak to you of things that you do not know.

But with God's army, there is no maximum time limit

of service, no discharge, full retirement and your benefits never end!

How safe do you feel in today's world?

Did you know that there is a level of supernatural personal protection that only Creator God can provide? "He that dwelleth in the secret place of the most High shall abide under the shadow of the Almighty ... For he shall give his angels charge over thee, to keep thee in all thy ways." (Psalm 91 KJV)

You may find this hard to believe, but it has been my personal experience over my lifetime that the safest possible place to be, no matter what or where, is not in a compound ... a fort ... an underground tunnel ... a cave ... but being where God directs you to be at the time! Even in places you would not normally trust!

When Creator God directs you He is also responsible for you! This your divine appointment!

What does this mean? It is your opportunity to choose the lifestyle you wish to have for the rest of your life and for eternity! A life of chances or whatevers or a life with love, purpose, meaning and freedom from all fear.

The Prayer of Connection:

"Please Holy Spirit of the Living God ... come into my life, guide and direct me from this moment on. I realize that I cannot earn this free gift from you. Any works of righteousness on my part do not even count. Forgive all of my shortcomings ... my missing of the mark of your high calling. I release my faith in the saving grace of Jesus Christ who paid this price for me with His death on the cross of Calvary. Come Holy Spirit and connect to my spirit being and live in me forever!"

Never forget ... God has already paid the price that you will never have to pay!

It amazes me how many people question whether God is good and if He is a loving God. Obviously, they have been told or taught that God is judgmental and critiques every step in life we make! That is what religion has done to people. In fact, there are several ministries that counsel thousands of people who have been abused by religion, legalistic churches and well-meaning Christians! What a shame. Religion and its rules simply do not demonstrate the true nature of Creator God!

What is the love of God like? If you have only been loved by another human being, you really do not know what being loved is all about. Loving as we may be we are all self-centered to one degree or another. Not with God. There is nothing He needs from you. But He has everything you need from Him. How does it feel? Someone once described it this way, "It's like feeling a gentle, warm, refreshing rain coming down on you on a sunny day!"

"God creates life in everything and new life in everything and everyone He touches!"

There is nothing else like it! He is a friend that is truer than a brother and He never leaves you nor forsakes you in any way, no matter what you may have done. It almost sounds too good to be true.

I have a photo on my office wall that has a picture of a beautiful dog. The caption reads, "God couldn't be physically with us so He gave us dogs. Notice that dog spelled backwards is God. And they both show unconditional love!" No ...I am not implying that God is a dog!

The purpose of God's plan is so that despite the chaos in life and world conditions, you have the peace and assurance that God is in control and His plans for you have a purpose. It begins with faith in Him. Then it grows to a deep trust. We depend too much on timetables but that is not as important as waiting on Him and for Him to bring everything to pass in our lives in the perfection of His time and creation.

It is wise to make your decision right now as there are anti-God spirit forces designed to distract you. Their goal is to

Peace is gazing at the stars with the knowledge that you know their creator.
©GodFruits.com

see to it that you do not accept this Connection with God! Don't let that happen to you. "For this shall everyone that is godly pray unto thee, in a time when thou mayest be found ..." (Psalm 32:6 KJV)

"And God even knows you by your name!"

Kirt G. Salisbury
San Diego, California

About The Author:

Kirt G. Salisbury was raised a small-town farm kid, was graduated from Engineering College, became Technical Director for ABC, served in the military in US Air Force-Okinawa, studied Marketing Communications at Harvard Business School, has worked as Para-Law Enforcement with Disaster Services, Biomedical Engineer and EMT, Professional Photojournalist, Ghost writer for major International Ministry Organizations, occupied positions in Executive Management, Bible College Graduate and President of Salisbury Alexander Communications.

In spite of this strange diverse background, there has been an orchestrated plan behind his life's path that has interconnected and intertwined for the basis of this book.

He discovered that "Things are not always as they appear!" Amazed at the degree of tunnel vision that exists today and the different opposing views and opinions on every imaginable subject from Science Theories, Religious Theology,

Military Strategies, Technological Adaptations, Politics, and Homeland Security ... he concluded that truth becomes a matter of whatever you can get someone to believe!

Who is right? Who is wrong? Are there any right answers? Can there truly be freedom, love and peace? What will transpire in the unknown future? What will mankind ultimately become? What will life ultimately become? Faced with these questions his search is the writ of this book. He reveals the CONNECTION that can provide these answers to anyone who desires them!

Kirt G. Salisbury
Salisbury Alexander Communications
San Diego, California